THE EMPTY CRADLE

The Global Population Bust, Tubal Fertility, and the Engineered Future of Humanity

THE EMPTY CRADLE

The Global Population Bust, Tubal Fertility, and the Engineered Future of Humanity

Mahendra Jagir

Studio of Books LLC
5900 Balcones Drive Suite 100
Austin, Texas 78731
www.studioofbooks.org
Hotline: (254) 800-1183

Ordering Information:
Special discounts are available on quantity purchases by corporations, associations, and others. For details, contact the publisher at the address above.

Printed in the United States of America.

ISBN-13: Paperback   978-1-970283-78-5
          Hardback   978-1-970283-79-2
          eBook   978-1-970283-80-8

Library of Congress Control Number: 2026909828

Dedication

To Dustin, Matthew, Georgia, and Jolie,

This work is written with a deep awareness of time—not just the time we measure in years, but the generational time that defines the continuity of humanity itself.

Every generation inherits a world shaped by the decisions of those who came before it. Some inherit stability. Others inherit uncertainty. And occasionally, a generation inherits a turning point—a moment when the direction of the future is no longer fixed, but must be actively determined.

You belong to such a generation.

The ideas contained in this book are not simply observations about society or projections about the future. They are reflections on a responsibility—the responsibility to ensure that the systems supporting human life continue to function, evolve, and endure.

Throughout history, continuity has often been taken for granted. The assumption that there will always be a next generation has rarely been questioned. Yet today, that assumption is no longer as certain as it once was.

This work is dedicated to you because the future it describes is not abstract. It is the world you will live in, shape, and ultimately pass forward.

May you inherit not a world in decline, but one that has recognized its challenges and chosen to address them with clarity, intention, and foresight.

May you inherit systems that are not only functional, but sustainable.

And may you carry forward the understanding that continuity is not automatic—it is created.

Preface

The Silent Clock

For much of modern history, the dominant narrative surrounding population has been one of excess.

The concern was not whether humanity would continue, but whether it would grow beyond the limits of its environment. The idea of overpopulation shaped economic theory, environmental policy, and global consciousness. It influenced how governments planned for the future, how institutions structured their systems, and how individuals understood their place within the world.

Growth was not only expected—it was feared.

This fear was not without reason. The twentieth century witnessed unprecedented increases in population. Advances in medicine reduced mortality rates. Improvements in sanitation extended life expectancy. Agricultural innovation expanded food production. These developments combined to produce a rapid expansion of human numbers.

This expansion appeared to confirm a long-standing assumption: that population growth, if left unchecked, would continue indefinitely.

Entire frameworks were built upon this assumption.

Economic systems relied on expanding labor forces. Urban planning anticipated increasing density. Educational institutions prepared for rising enrollment. Environmental discourse focused on managing the consequences of growth.

The narrative was consistent.

Humanity's greatest demographic challenge was too many people.

Yet, while attention remained fixed on this concern, a different process began to unfold.

Quietly, gradually, and without widespread recognition, the trajectory began to change.

Birth rates declined.

At first, this decline was interpreted as part of a natural progression. The demographic transition model suggested that as societies developed, fertility would decrease. Families would choose to have fewer children as survival rates improved and economic structures shifted.

This transition was understood as stabilization.

Growth would slow. Populations would reach equilibrium.

However, in many regions, the decline did not stop at equilibrium.

Fertility rates continued to fall.

What was expected to stabilize instead began to contract.

This shift did not occur suddenly.

It unfolded over decades, embedded within statistical reports, demographic studies, and everyday observation. It appeared in subtle ways: fewer children in classrooms, aging communities, declining birth announcements.

Individually, these changes did not appear alarming.

Collectively, they revealed a pattern.

The pattern indicated a reversal.

The direction of population growth—long assumed to move forward—had begun to shift.

This reversal is significant not because it is immediate, but because it is cumulative.

Unlike crises that emerge suddenly, the Population Bust operates through accumulation. Its effects build gradually, often remaining unnoticed until they reach a threshold at which they can no longer be ignored.

This gradual nature makes it difficult to address.

Societies are structured to respond to visible disruption. They are less equipped to recognize slow transformation. When change occurs incrementally, it can be absorbed without triggering a sense of urgency.

Yet the absence of urgency does not imply the absence of consequence.

At its core, the Population Bust represents a shift in the conditions under which human life continues.

It is not driven by external catastrophe.

It is driven by internal evolution.

The systems that have improved quality of life—economic development, technological advancement, expanded individual autonomy—have also altered the environment in which decisions about reproduction are made.

In earlier periods, reproduction was embedded within necessity.

Children contributed to labor. They ensured continuity within family structures. They were integrated into the economic and social fabric of daily life.

In such contexts, reproduction was not only natural.

It was expected.

Modern systems have changed these conditions.

Introduction

The Great Quiet

There is a difference between noise and signal.

Noise is immediate. It disrupts, demands attention, and compels response. It is visible, measurable, and often urgent.

Signal, by contrast, is subtle.

It unfolds gradually, requiring observation rather than reaction. It does not announce itself. It accumulates over time, becoming visible only when patterns are examined across broader scales.

The transformation described in this book belongs to the latter category.

It is not marked by a singular event.

It is defined by absence.

The absence of births.

This absence does not generate headlines. It does not produce immediate disruption. Instead, it manifests quietly, embedded within data, trends, and lived experience.

Over time, this absence becomes a pattern.

That pattern may be described as the Great Quiet.

In communities around the world, the indicators of this quiet are present.

Schools close or consolidate due to declining enrollment. Neighborhoods age without renewal. Public spaces reflect shifting demographics. The proportion of older individuals increases while the number of younger individuals declines.

These changes are not isolated.

They are interconnected.

They form part of a broader transformation in population dynamics.

At the center of this transformation lies a fundamental principle.

For a population to remain stable, it must replace itself.

This principle is quantified by the replacement rate—approximately 2.1 children per woman.

When fertility falls below this level, the long-term trajectory becomes one of decline.

In many regions, this threshold has already been crossed.

The implications of this shift are not immediately apparent.

Population momentum sustains overall numbers for extended periods. Improvements in healthcare extend life expectancy, allowing populations to remain stable or even grow temporarily.

However, this stability is temporary.

The underlying trend persists.

Over time, the effects become more pronounced.

The age structure of society shifts. The proportion of individuals in the workforce declines relative to those who depend on it. Economic systems adjust, often under strain.

These changes occur gradually.

They do not produce immediate disruption.

Yet their cumulative impact is significant.

Understanding the Great Quiet requires examining its causes.

These causes are complex.

They do not originate from a single factor.

They emerge from the interaction of multiple systems.

Economic conditions influence decisions about family formation. The cost of housing, childcare, and living expenses acts as a constraint.

Social structures shape life trajectories. Education and career development extend into early adulthood, delaying family formation.

Biological systems, however, remain constant.

This creates a misalignment.

Psychological factors further influence decision-making. Perceptions of stability, security, and future conditions shape long-term commitments.

In an environment characterized by uncertainty, such commitments may be delayed.

Taken together, these factors form a system.

The decline in birth rates is not the result of individual preference alone.

It is the outcome of systemic interaction.

This complexity makes the issue difficult to address.

Traditional responses focus on isolated variables—economic incentives, policy changes, or cultural messaging.

While these approaches may influence behavior, they do not resolve the underlying system.

This suggests that a broader perspective is required.

An approach that considers not only the causes of decline, but the conditions necessary for continuity.

The chapters that follow explore these conditions.

They examine the economic, social, and psychological factors that contribute to the Population Bust. They analyze the consequences of sustained decline. They consider the potential for technological and systemic adaptation.

The goal is not to prescribe a single solution.

It is to expand the framework within which solutions are considered.

The Great Quiet is not an endpoint.

It is a transition.

What follows will determine whether that transition leads to adaptation or continued decline.

Table of Contents

Dedication i

Preface iii

Introduction vii

PART I The Diagnosis

CHAPTER 1
The Apex – The Silent Turning Point 1

CHAPTER 2
Economic Sterilization – The Cost of Existence 11

CHAPTER 3
The Social Shift – The Structural Impasse 19

PART II The Consequences

CHAPTER 4
The Inverted Pyramid – The Global Breakdown 31

PART III The Remedy

CHAPTER 5
The Machine Mother – The Tubal Revolution 43

CHAPTER 6
Sustainability and the Engineered Future 51

CHAPTER 7
The Choice – The Point of Divergence 61

Conclusion 73

Afterwords 81

Author's Note 85

Appendices 87

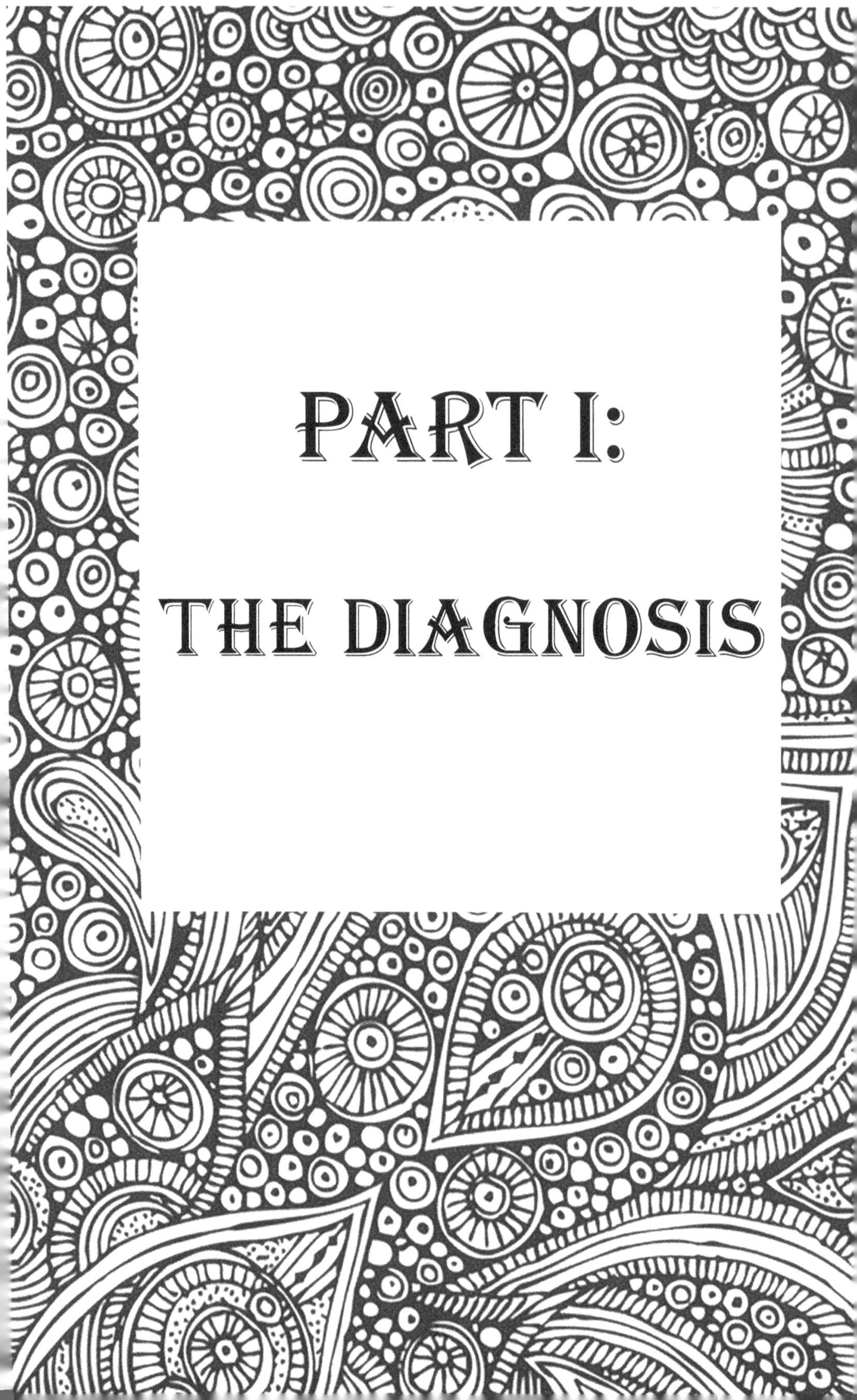

PART I:

THE DIAGNOSIS

CHAPTER 1

The Apex – The Silent Turning Point

For most of human history, population growth was not merely observed—it was assumed.

It existed as a constant, embedded so deeply within the structure of life that it required no explanation. Generations followed one another in a continuous sequence, sustained by biological processes and reinforced by economic and social systems. The expectation of continuity was not questioned because it had never been meaningfully disrupted.

Growth signified vitality.

It represented labor, expansion, and the persistence of human presence across time. The continuation of the population was both a biological outcome and a structural necessity. Societies depended on it, and in turn, societies reinforced it.

This relationship between population and structure defined the development of civilization.

Agricultural systems relied on family labor. Industrial systems depended on expanding workforces. Urban environments were designed with the expectation of increasing density. Cultural narratives emphasized lineage, inheritance, and generational continuity.

In this context, population growth was not a variable.
It was a foundation.

The Historical Assumption of Continuity

To understand the significance of the present shift, it is necessary to examine the strength of the assumption it challenges.

For thousands of years, human populations were shaped by high birth rates and high mortality rates. Families had multiple children, not only as a cultural norm, but as a practical necessity. Survival was uncertain, and continuity required redundancy.

Children were integrated into the economic structure of the household. They contributed to labor, supported aging family members, and ensured the transmission of knowledge and resources.

Reproduction, therefore, was not merely biological.

It was structural.

Even as mortality rates declined and societies became more stable, this structure persisted. The expectation that each generation would replace and expand upon the previous one remained intact.

This expectation became embedded in institutions.

Education systems were designed to accommodate growing populations. Economic models assumed expanding labor markets. Governments planned for future growth in infrastructure and services.

Continuity was not only assumed.

It was planned.

The Rise of the Overpopulation Paradigm

By the twentieth century, this assumption began to produce concern.

Advances in medicine, sanitation, and agricultural production led to rapid population growth. Mortality rates declined while birth rates remained relatively high. The result was an acceleration in population expansion that appeared unprecedented.

This acceleration gave rise to a new narrative.

Overpopulation.

The concern was no longer whether humanity would continue, but whether it would exceed the capacity of its environment. The idea that population growth could outpace resources became central to global discourse.

This narrative influenced policy, research, and public perception.

Efforts were made to manage population growth through education, access to contraception, and policy interventions. The goal was to prevent expansion from becoming unsustainable.

The assumption, however, remained intact.

Growth was still expected.

The question was how to limit it.

The Quiet Emergence of Decline

While attention remained focused on managing growth, a different process began to unfold.

Gradually, and without dramatic visibility, birth rates began to decline.

At first, this decline appeared consistent with the demographic transition model. As societies developed, fertility rates decreased. Families chose to have fewer children as survival rates improved and economic structures shifted.

This trend was interpreted as stabilization.

Population growth would slow and eventually reach equilibrium.

However, in many regions, the decline did not stop at equilibrium.

Fertility rates continued to fall, moving below the level required to sustain the population.

This marked the beginning of a new phase.

A phase not of growth, but of contraction.

Defining the Apex

This turning point may be described as the Apex.

The Apex is not a singular event.

It is a phase—a transition point at which the trajectory of population growth shifts from expansion to decline.

It is the moment at which continuity is no longer self-sustaining.

This definition is both simple and profound.

For the first time in modern history, the systems that once ensured continuity no longer produce it automatically.

The Apex is not defined by visible disruption.

It is defined by underlying change.

It exists within data, within trends, and within the cumulative outcomes of individual decisions.

The Replacement Threshold and Its Implications

At the center of the Apex lies a fundamental metric: the replacement rate.

To maintain a stable population, each generation must produce enough offspring to replace itself. This rate is commonly estimated at approximately 2.1 children per woman.

When fertility remains at or above this level, populations sustain themselves.

When it falls below, decline becomes inevitable.

This decline is not immediate.

Population momentum can sustain overall numbers for extended periods. Larger generations from previous decades continue to contribute to total population figures, masking the decline in births.

However, this momentum is temporary.

Over time, the absence of sufficient replacement becomes evident. The population structure shifts. The proportion of older individuals increases relative to younger ones.

The consequences accumulate.

The Illusion of Stability

One of the defining characteristics of the Apex is the illusion it creates.

At a surface level, population figures may appear stable. In some cases, they may even continue to grow. This creates the perception that continuity remains intact.

However, this perception is misleading.

The underlying structure tells a different story.

Instead of a broad base of younger individuals, populations begin to exhibit a narrowing base. The age distribution shifts. The balance between generations changes.

This structural shift does not produce immediate disruption.

It unfolds gradually.

This gradual nature allows it to remain largely unrecognized.

From Biological Default to Conscious Decision

The most significant transformation underlying the Apex is the shift from reproduction as a default outcome to reproduction as a conscious decision.

In earlier systems, reproduction was embedded within life.

Individuals formed partnerships, and children followed as a natural progression. The process required little deliberation because it was integrated into the structure of society.

Modern systems have altered this integration.

Reproduction is no longer automatic.

It is optional.

This optionality introduces evaluation.

Individuals assess whether to have children based on economic conditions, personal aspirations, and perceptions of the future.

Each of these factors introduces complexity.

When reproduction becomes subject to evaluation, it becomes subject to delay.

Delay, when aggregated across a population, produces measurable effects.

Behavioral Shifts in Conditions of Abundance

One framework that offers insight into this transformation is the concept of the Behavioral Sink.

In controlled environments where basic needs are met and external pressures are minimized, populations do not necessarily continue to grow. Instead, they may exhibit changes in behavior that reduce reproduction.

This phenomenon challenges intuitive assumptions.

Abundance does not guarantee growth.

While human societies are more complex than experimental environments, certain parallels can be observed.

Modern societies provide unprecedented access to resources, information, and opportunity. At the same time, they introduce new forms of pressure—competition, overstimulation, and fragmentation.

These conditions influence behavior.

Reproduction becomes less central.

Other priorities emerge.

The Fragmentation of Continuity

The Apex represents more than a demographic threshold.

It reflects a fragmentation of continuity.

Continuity, in this context, refers to the seamless progression from one generation to the next. It is the process by which populations sustain themselves without conscious intervention.

In modern society, this process has been disrupted.

Reproduction is no longer integrated into the default trajectory of life.

It exists as one option among many.

This fragmentation introduces uncertainty.

Continuity becomes conditional.

The Systemic Nature of the Shift

It is important to recognize that the Apex is not the result of individual choices alone.

It is a systemic condition.

Individuals make decisions within the context of their environment. When that environment changes, the outcomes of those decisions change.

Economic structures, social norms, technological advancements, and psychological conditions all contribute to this environment.

The decline in fertility is the cumulative result of these influences.

Understanding the Apex requires examining the system as a whole.

A Turning Point Without Announcement

Unlike historical turning points marked by visible events, the Apex occurs without announcement.

There is no singular moment at which it becomes apparent.

It emerges through accumulation.

Over time, patterns become clear.

Data reflects the trend. Observations confirm it. The structure of society begins to shift.

Yet because the process is gradual, it does not trigger immediate response.

This is both its defining characteristic and its greatest challenge.

The Implications of Recognition

Recognizing the Apex changes the framework within which population dynamics are understood.

If continuity is no longer automatic, it must be supported.

This recognition shifts the focus from assumption to intention.

The question is no longer whether population will continue to grow.

It is whether it will continue at all, and under what conditions.

The Beginning of a New Inquiry

The Apex marks the beginning of a broader inquiry.

It raises questions about the systems that support reproduction, the conditions under which individuals make decisions, and the potential pathways toward stability.

These questions extend beyond demographics.

They touch on the structure of society itself.

The chapters that follow explore these dimensions.

They examine the economic, social, and psychological factors that contribute to the Population Bust. They analyze the consequences of sustained decline. And they consider the possibilities for adaptation.

The Silent Turning Point

The Apex is silent.

It does not announce itself through crisis or disruption.

It reveals itself through absence.

The absence of births.

The absence of replacement.

The absence of continuity.

Yet within this silence lies significance.

It marks the point at which the trajectory of human population changes.

It marks the beginning of a new phase.

A phase in which continuity is no longer guaranteed.

A phase in which the future must be understood differently.

Toward Understanding and Design

From this point forward, population dynamics cannot be treated as a constant.

They must be examined, understood, and, where necessary, supported.

The Apex is not an endpoint.

It is a starting point.

It is the moment at which humanity becomes aware of a shift that has already begun.

What follows will determine how that shift unfolds.

CHAPTER 2

Economic Sterilization – The Cost of Existence

If the Apex represents the moment at which population growth ceases to sustain itself, then the forces driving that shift must be understood in detail.

Among these forces, economic conditions occupy a central position.

While biological capacity remains largely unchanged, the environment in which reproductive decisions are made has evolved significantly. This evolution has altered not only the feasibility of having children, but the perception of what is required to do so.

The concept of Economic Sterilization captures this transformation.

It does not refer to a loss of biological ability.

It refers to a system in which economic structures function as constraints on reproduction.

Individuals may retain both the capacity and, in many cases, the desire to have children. However, the conditions necessary to support that decision have become increasingly difficult to attain.

The result is not explicit prohibition.

It is implicit limitation.

From Survival Economy to Optimization Economy

To understand Economic Sterilization, it is necessary to examine the transformation of economic systems.

In earlier periods, economic life was organized around survival.

Resources were limited. Labor was essential. Families operated as economic units in which each member contributed to collective stability. Children were integrated into this structure, providing both immediate and future value.

Reproduction, therefore, aligned with economic necessity.

More children meant greater labor capacity and increased security.

Modern economies operate differently.

They are organized around optimization rather than survival.

Productivity, efficiency, and specialization define economic participation. Individuals invest in education and training to increase their value within the system. Economic success is tied to skill development, mobility, and long-term planning.

In this context, children are no longer economic assets in the traditional sense.

They are investments.

This shift changes the relationship between reproduction and economics.

The Rising Cost of Stability

The transition from survival to optimization has increased the cost of stability.

In earlier systems, stability was achieved through participation in established structures. Employment was often localized, housing was relatively accessible, and the transition to independence occurred earlier.

In modern systems, stability requires extended investment.

Education spans many years. Career development requires sustained effort. Housing costs have risen significantly, often outpacing income growth.

This increase in cost affects decision-making.

Parenthood is not considered in isolation.

It is evaluated within the broader context of economic readiness.

The threshold for readiness has risen.

Housing as Structural Barrier

Housing represents one of the most significant components of Economic Sterilization.

It is not merely a physical requirement.

It is a foundation for stability.

In many regions, the cost of housing has increased beyond the reach of a large portion of the population. Property ownership, once considered a standard milestone, has become increasingly difficult to attain.

Rental markets, while providing alternatives, often impose significant financial burden.

This burden affects long-term planning.

Without stable housing, the prospect of raising children becomes uncertain.

Housing influences not only feasibility, but perception.

It shapes how individuals assess their readiness for parenthood.

The Expansion of Child-Rearing Costs

The cost of raising a child has increased in both scope and expectation.

Modern standards include not only basic necessities, but also education, healthcare, and developmental support.

These standards reflect improvements in quality of life.

However, they also raise the financial threshold required for parenthood.

Childcare costs are particularly significant.

In many regions, childcare expenses constitute a substantial portion of household income. For dual-income households, which are often necessary to meet overall expenses, childcare is not optional.

This creates structural tension.

Income is required to support children, yet the cost of enabling that income reduces its effectiveness.

In some cases, the cost of childcare approaches or exceeds the income of one parent.

This dynamic influences decisions.

The Delay of Financial Independence

Modern economic systems extend the transition into adulthood.

Education and training delay entry into the workforce. Career establishment requires time. Financial independence is achieved later.

This delay has cascading effects.

Partnership formation is postponed. Family formation is deferred. The window for reproduction narrows.

This process is often framed as choice.

However, it is deeply influenced by economic structure.

The timing of stability determines the timing of reproduction.

Employment Precarity and Risk

The nature of employment has evolved.

Modern labor markets prioritize flexibility and adaptability. While these characteristics create opportunity, they also introduce instability.

Employment may be less secure. Income may fluctuate. Career paths may be uncertain.

This condition, often described as precarity, affects long-term planning.

Parenthood requires stability.

It involves commitment over extended periods.

When employment conditions are uncertain, the willingness to undertake such commitments decreases.

This response is rational.

The Psychological Weight of Economic Pressure

Economic conditions influence not only material reality, but perception.

Financial uncertainty creates psychological pressure.

This pressure affects decision-making.

Individuals prioritize immediate stability over long-term commitments. Parenthood, which requires sustained investment, becomes more difficult to justify.

This dynamic reinforces structural constraints.

The decision to delay or avoid parenthood becomes a means of managing uncertainty.

The Rising Standard of Readiness

In addition to structural constraints, societal expectations have evolved.

The standard for readiness has increased.

Individuals often feel that they must achieve financial security, career stability, and personal development before considering children.

These expectations reflect a desire to provide optimal conditions.

However, they also raise the threshold for entry.

As the threshold rises, fewer individuals meet it within the available timeframe.

The Economics of Opportunity Cost

Parenthood involves opportunity cost.

Time and resources allocated to raising children cannot be allocated elsewhere.

In modern systems, this cost is significant.

Career progression may be interrupted. Income potential may be reduced. Opportunities for mobility may be limited.

These trade-offs influence decision-making.

Individuals evaluate the cost of parenthood relative to other opportunities.

In many cases, the cost is perceived as high.

The Dual-Income Dependency

Modern households often depend on dual incomes.

This dependency reflects the cost structure of contemporary life.

Housing, childcare, and living expenses require combined earnings.

However, dual-income structures introduce complexity.

Balancing work and caregiving responsibilities creates strain.

The cost of maintaining this balance influences decisions about family size and timing.

The Feedback Loop of Economic Sterilization

Economic Sterilization operates through feedback loops.

As birth rates decline, demographic structures shift. This shift influences economic conditions, creating additional constraints.

A declining workforce affects economic growth. Reduced growth limits opportunities for younger generations. These limitations reinforce the conditions that contribute to delayed reproduction.

This cycle is self-reinforcing.

Systemic Nature of Economic Constraint

Economic Sterilization is not the result of a single factor.

It is a system.

Housing costs, childcare expenses, employment conditions, and societal expectations interact to create an environment in which reproduction is constrained.

Understanding this system is essential.

It cannot be addressed through isolated measures.

Beyond Incentives

Traditional approaches focus on incentives.

Financial support and policy adjustments aim to reduce the cost of parenthood.

While these measures may provide relief, they do not address the underlying structure.

Economic Sterilization is not solely a matter of cost.

It is a matter of alignment.

The Intersection with Social Systems

Economic conditions intersect with social and psychological factors.

The delay in financial stability influences relationships. The cost of living shapes expectations. Employment conditions affect long-term planning.

These interactions create complexity.

Constraint on Continuity

At its core, Economic Sterilization represents a constraint on continuity.

The ability to sustain the population depends on the conditions that support reproduction.

When those conditions become restrictive, continuity is affected.

CHAPTER 3

The Social Shift – The Structural Impasse

If economic systems define the material constraints within which individuals operate, social systems define how those constraints are interpreted, navigated, and ultimately internalized.

They shape not only what individuals can do, but what they believe they should do.

They determine how time is structured, how identity is formed, and how meaning is assigned to life decisions.

It is within this domain that one of the most profound transformations of modern civilization has occurred.

This transformation is not immediately visible.

It is not marked by a single event or a clear point of rupture.

It is gradual, cumulative, and deeply embedded.

Yet its effects are unmistakable.

It has altered the structure of life.

It has redefined the trajectory of adulthood.

And most critically, it has disrupted the continuity of reproduction.

This disruption may be understood as the Structural Impasse.

The Collapse of Default

For most of human history, life followed a pattern that required little explanation.

It was not rigid in every detail, but it was consistent in its sequence.

Individuals transitioned from childhood to adulthood, formed partnerships, and had children. This sequence was not merely encouraged—it was embedded within the structure of society.

It was the default.

This default was reinforced by necessity.

Economic systems depended on family units. Social structures supported continuity. Cultural expectations aligned with reproduction.

The result was a form of structural momentum.

Life progressed in a direction that required minimal deliberation.

Reproduction was not a decision.

It was a consequence.

Modern society has dismantled this default.

The pathways that once guided individuals toward family formation have been replaced by open frameworks.

Education extends into early adulthood. Career development requires sustained investment. Personal identity becomes an ongoing process rather than a fixed outcome.

In this environment, reproduction is no longer embedded.

It is optional.

This shift—from default to option—is foundational.

When an action is no longer default, it becomes subject to evaluation.

Evaluation introduces hesitation.

Hesitation introduces delay.

And delay, when aggregated across populations, produces decline.

The Expansion of Choice and the Burden of Decision

The expansion of individual choice is often regarded as one of the defining achievements of modern society.

Individuals are no longer constrained by rigid roles or predetermined paths. They have access to education, mobility, and opportunities that were historically unavailable.

This expansion has increased freedom.

It has also increased complexity.

Choice requires decision-making.

Decision-making requires evaluation.

Evaluation introduces uncertainty.

In earlier systems, reproduction occurred within a limited set of possibilities.

In modern systems, it competes with a vast array of alternatives.

Career advancement, personal development, travel, creative pursuits, and lifestyle preferences all occupy time and attention.

Reproduction must be considered within this broader landscape.

It is no longer assumed.

It must be chosen.

This transformation introduces a cognitive dimension.

Individuals must weigh costs and benefits, assess timing, and anticipate long-term consequences.

The result is not simply choice.

It is decision burden.

Time as a Structural Constraint

Time has become one of the most significant variables in modern life.

It is no longer structured primarily by biological or social rhythms.

It is structured by systems.

Education requires extended periods of preparation. Careers demand sustained investment. Economic stability often arrives later than in previous generations.

This restructuring of time creates conflict.

The biological window for reproduction remains fixed.

The social timeline has expanded.

This misalignment produces compression.

Multiple life stages must occur within a limited biological timeframe.

Education, career establishment, partnership formation, and child-rearing must be negotiated within overlapping periods.

This compression increases complexity.

It reduces flexibility.

And it amplifies the consequences of delay.

The Feminine Pivot – Expansion and Constraint

At the center of the Social Shift lies the Feminine Pivot.

The integration of women into education, professional life, and leadership represents one of the most significant transformations in modern history.

It has expanded the productive capacity of society.

It has increased intellectual diversity and innovation.

It has redefined expectations across all domains.

However, this transformation intersects with biological constraints.

The timeline for achieving professional stability has extended.

The biological window for reproduction has not.

This creates structural tension.

The decision to pursue education and career development often coincides with the period of peak fertility.

This overlap forces trade-offs.

Career progression requires continuity.

Parenthood introduces interruption.

The cost of interruption is not evenly distributed.

It affects income, opportunity, and long-term trajectory.

This dynamic leads to delay.

Delay reduces reproductive capacity.

The result is not simply a personal trade-off.

It is a systemic constraint.

Structural Asymmetry and the Persistence of Burden

Despite advances in equality, certain asymmetries remain embedded within biological reality.

Pregnancy imposes physical demands that cannot be fully redistributed.

These demands intersect with social expectations.

The concept of the "double burden" reflects this intersection.

Individuals—particularly women—are expected to maintain full participation in professional life while also assuming primary responsibility for caregiving.

This expectation creates strain.

The effort required to sustain dual roles influences decision-making.

Parenthood becomes not only a personal commitment, but a structural challenge.

In this context, delay becomes rational.

Reduction becomes likely.

The Delay of Union and the Compression of Intimacy

The timing of relationships has shifted significantly.

Partnership formation occurs later in life.

This delay reflects extended education, career development, and evolving social expectations.

While this shift increases individual preparedness, it reduces available time.

Biological systems operate within defined limits.

As partnerships form later, the window for reproduction narrows.

The number of potential births decreases.

This dynamic produces a gap between intention and outcome.

Individuals may desire children.

Yet they may not achieve intended family size.

This gap is critical.

It demonstrates that declining fertility is not solely a matter of preference.

It is a consequence of structural timing.

The Individualization of Identity

Modern society emphasizes individual identity.

Personal fulfillment, autonomy, and self-expression are central values.

These values reshape how individuals approach life decisions.

Parenthood is no longer the defining feature of adulthood.

It is one option among many.

This diversification reflects progress.

However, it introduces competition.

Time, energy, and resources must be allocated across multiple domains.

Reproduction competes with other forms of fulfillment.

Career, creativity, exploration, and personal growth all occupy space within the modern life structure.

The decision to have children becomes contingent.

It requires alignment of multiple conditions.

The Fragmentation of Community and Support

Traditional support systems have weakened.

Extended families are less central.

Geographic mobility separates individuals from established networks.

Communities are more fluid and less stable.

This fragmentation increases the burden of parenthood.

Responsibilities that were once distributed across networks are now concentrated within smaller units.

The absence of support amplifies both perceived and actual difficulty.

The concept of "Homelessness of the Spirit" reflects this condition.

It describes a lack of stable belonging.

Without enduring support structures, long-term commitments become more difficult to sustain.

Psychological Overload and Decision Paralysis

The expansion of choice introduces psychological complexity.

Multiple variables must be considered.

Each variable introduces uncertainty.

The cumulative effect is overload.

Decision-making becomes more difficult.

In such environments, delay becomes a default response.

This delay is not necessarily conscious.

It emerges from the structure of the decision-making process.

Over time, delay leads to reduced outcomes.

The Spectrum Paradox – Awareness and System Capacity

The increased visibility of neurodivergent conditions introduces additional complexity into reproductive decision-making.

Awareness has improved understanding.

It has also introduced new considerations.

The possibility of long-term caregiving responsibility becomes part of the evaluation.

The concept of the "caregiver trap" reflects this concern.

It describes extended responsibility in the absence of systemic support.

This concern is not rooted in bias.

It is rooted in system capacity.

When support systems are insufficient, perceived burden increases.

This perception influences decision-making.

Convergence into Structural Impasse

All elements converge.

Choice expands.

Time compresses.

Support systems fragment.

Psychological complexity increases.

These factors interact.

They form a system.

Within this system, reproduction is no longer seamlessly integrated.

It is constrained.

This is the Structural Impasse.

The Impasse as Systemic Outcome

The Structural Impasse is not the result of individual decisions alone.

It is the outcome of systemic interaction.

Economic, social, and psychological factors reinforce one another.

The decline in fertility emerges from this interaction.

Understanding this requires a systemic perspective.

The Need for Realignment

Addressing the Structural Impasse requires realignment.

Systems must be adjusted to support continuity.

This does not require reducing autonomy.

It requires aligning autonomy with sustainability.

Transition to Consequence

The Structural Impasse produces consequences.

These consequences extend beyond individual behavior.

They reshape society.

The next chapter examines this transformation.

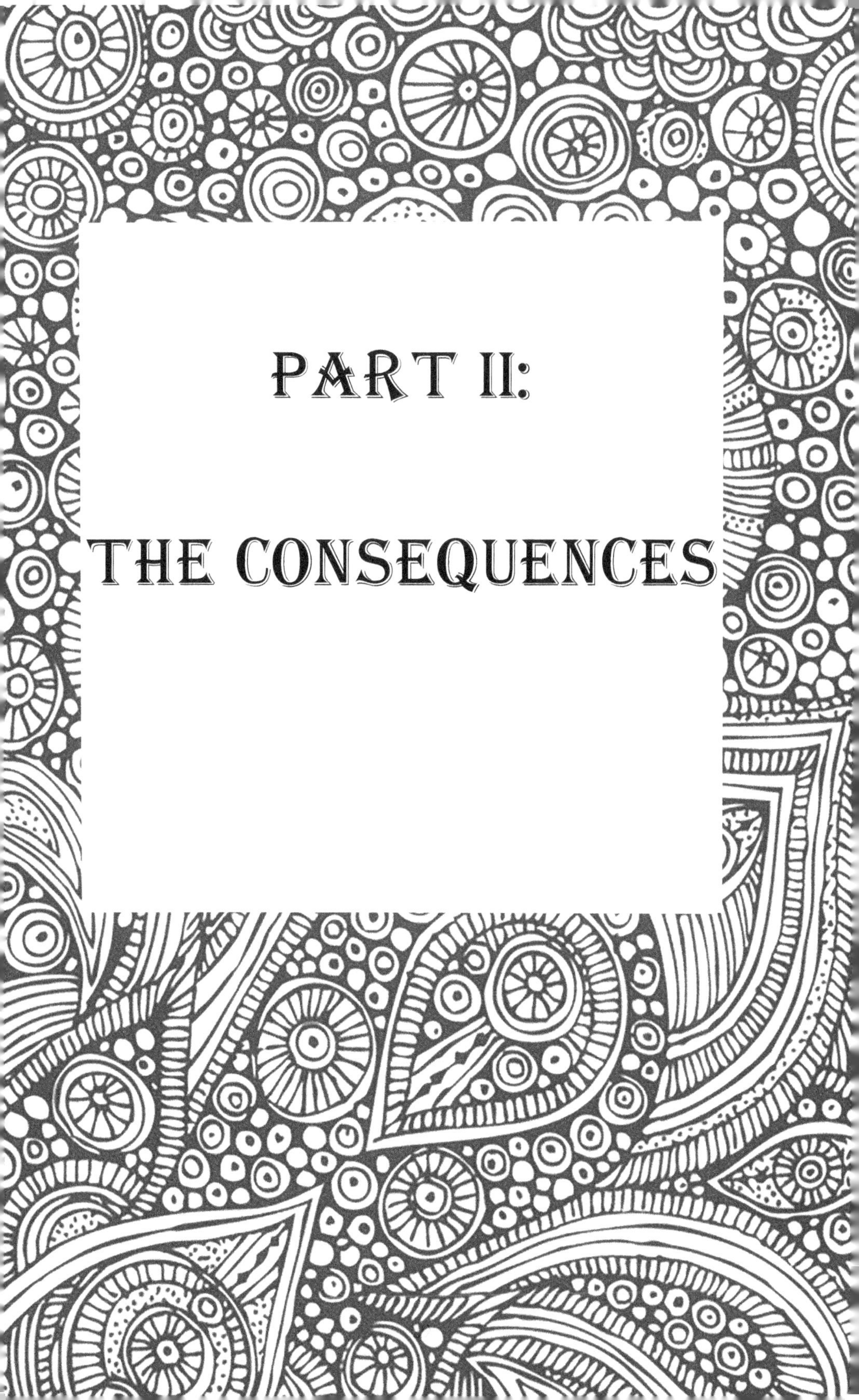
PART II:

THE CONSEQUENCES

CHAPTER 4

The Inverted Pyramid – The Global Breakdown

The Structural Impasse described in the previous chapter does not remain contained within individual lives.

It scales.

What begins as delayed decisions, reduced family size, and altered life trajectories accumulates across populations. Over time, these individual outcomes converge into a structural transformation—one that reshapes the architecture of society itself.

This transformation is demographic.

But its consequences extend far beyond population numbers.

They reach into the foundations of economic systems, infrastructure design, cultural dynamics, and long-term sustainability.

The traditional population pyramid—long regarded as a stable and self-sustaining structure—is no longer intact.

It is inverting.

The Population Pyramid as Civilizational Blueprint

For generations, the population pyramid has served as an implicit blueprint for civilization.

At its base lies a broad population of younger individuals—children and working-age adults—whose productivity supports the economic and social systems above. At its top lies a smaller population of older individuals who, having exited the workforce, depend on those below.

This structure creates balance.

Each generation supports the next.

The system sustains itself.

Economic growth, infrastructure expansion, and social stability all depend on this balance.

It is not merely demographic.

It is structural.

The Mechanics of Inversion

The inversion of the population pyramid occurs through two simultaneous processes.

First, fertility declines.

Fewer individuals are born into each successive generation.

Second, longevity increases.

Individuals live longer, extending the duration of life beyond previous norms.

These processes interact.

The base of the pyramid narrows.

The top expands.

Over time, the structure reverses.

The proportion of older individuals increases relative to younger ones.

This shift is gradual.

It unfolds over decades.

Yet its effects are cumulative.

The Illusion of Continuity

One of the most deceptive aspects of demographic inversion is the illusion it creates.

At a surface level, populations may appear stable.

Total numbers may remain constant or even increase due to population momentum.

This creates the perception that continuity remains intact.

However, this perception is misleading.

The underlying structure tells a different story.

The capacity for renewal is diminishing.

The illusion of continuity delays recognition.

And delayed recognition delays response.

The Dependency Ratio and Systemic Pressure

At the center of demographic inversion lies the dependency ratio.

This ratio reflects the relationship between those who are economically active and those who depend on them.

In a stable system, multiple workers support each retiree.

This balance allows economic and social systems to function effectively.

As the population ages, this ratio shifts.

Fewer workers are available to support more retirees.

This shift introduces systemic pressure.

Healthcare systems must accommodate increased demand.

Pension systems must extend benefits over longer periods.

Public resources must be reallocated.

Each adjustment carries cost.

Over time, the cumulative burden increases.

The Silver Tsunami as Structural Momentum

The aging population has often been described as the "Silver Tsunami."

This term captures both its scale and inevitability.

Unlike sudden disruptions, this transformation unfolds gradually.

It is not a wave that crashes.

It is a tide that rises.

Each year, the proportion of older individuals increases.

Each year, the system absorbs additional pressure.

Individually, these changes may appear manageable.

Collectively, they create momentum.

This momentum is difficult to reverse.

Workforce Contraction and Economic Recalibration

A declining birth rate leads to a contraction of the workforce.

Fewer individuals enter the labor market.

This affects productivity, economic growth, and systemic sustainability.

The effects are not immediate.

They emerge over time as smaller generations replace larger ones.

However, once the transition begins, it accelerates.

Labor shortages appear in key sectors.

Economic output slows.

The capacity to maintain systems diminishes.

This contraction interacts with aging.

As more resources are directed toward supporting older populations, fewer resources remain for investment.

This creates tension within economic systems.

The Shift from Expansion to Maintenance

Modern economies are structured for growth.

They rely on expansion.

Increasing populations drive consumption, innovation, and investment.

When growth slows or reverses, these assumptions no longer hold.

The system must adapt.

The focus shifts from expansion to maintenance.

Maintaining infrastructure, services, and economic stability becomes the priority.

This shift alters economic behavior.

Risk tolerance decreases.

Innovation slows.

Investment becomes more conservative.

Infrastructure Misalignment

Infrastructure is designed with assumptions about population.

Transportation networks, housing developments, utilities, and public services are built to accommodate growth.

When growth slows or reverses, these systems become misaligned.

Infrastructure does not contract easily.

Roads must still be maintained.

Utilities must remain operational.

Public institutions must continue to function.

However, the population supporting these systems declines.

This creates inefficiency.

Costs remain constant or increase.

Revenue decreases.

The result is structural imbalance.

The Emergence of Hollowed Systems

As populations decline, systems become hollow.

Schools operate below capacity.

Hospitals serve aging populations with fewer contributors.

Public services struggle to maintain coverage.

This hollowing is gradual.

It does not produce immediate collapse.

Instead, it produces degradation.

Systems function, but less efficiently.

Over time, this degradation accumulates.

Regional Decline and Spatial Imbalance

Demographic inversion does not occur uniformly.

Some regions experience rapid decline.

Others remain stable or continue to grow.

This creates spatial imbalance.

Declining regions may experience economic contraction, reduced services, and population loss.

Growing regions may face different challenges, such as resource allocation and infrastructure strain.

The divergence between regions introduces complexity.

Migration partially redistributes population.

However, it does not resolve structural differences.

The Feedback Loop of Decline

Demographic inversion operates through feedback loops.

Population decline reduces economic activity.

Reduced economic activity limits opportunity.

Limited opportunity accelerates population decline.

This cycle is self-reinforcing.

Each stage amplifies the next.

Breaking this cycle requires systemic intervention.

Innovation and Generational Energy

Innovation is closely tied to demographic structure.

Younger populations tend to drive experimentation, risk-taking, and adaptation.

As populations age, this dynamic shifts.

Older populations prioritize stability.

This shift influences the pace of innovation.

With fewer younger individuals, the capacity for rapid change decreases.

The system becomes more conservative.

Cultural Transformation and Temporal Orientation

Demographic inversion also affects culture.

Societies orient themselves around their dominant age groups.

Younger societies emphasize growth, experimentation, and future expansion.

Older societies emphasize preservation, stability, and risk management.

As populations age, cultural orientation shifts.

This shift influences policy, economic behavior, and social values.

The future becomes less expansive.

It becomes more cautious.

Global Asymmetry and Systemic Imbalance

The Population Bust is not global in uniformity.

Some regions experience declining fertility.

Others maintain higher birth rates.

This creates global asymmetry.

Regions with declining populations may face economic contraction.

Regions with growing populations may face resource challenges.

Migration attempts to balance these differences.

However, it introduces its own complexities.

Cultural, political, and economic factors influence its effectiveness.

Interconnected Systems and Amplification

The effects of demographic inversion are interconnected.

Changes in one domain influence others.

Economic shifts affect infrastructure.

Infrastructure affects quality of life.

Quality of life influences population decisions.

This interconnectedness amplifies the impact.

What begins as demographic change becomes systemic transformation.

The Breakdown of Growth Assumptions

Modern systems assume growth.

When growth ceases, these systems encounter limitations.

Economic models must adapt.

Infrastructure must be re-evaluated.

Policies must be reconsidered.

This requires a shift in perspective.

Growth is no longer the default.

The Threshold of Structural Change

As inversion progresses, societies approach a threshold.

At this threshold, incremental adjustments are insufficient.

Structural change becomes necessary.

This change may involve new economic models, technological integration, and policy innovation.

The form of adaptation will vary.

The need for adaptation is universal.

The Limits of Conventional Solutions

Traditional responses focus on increasing birth rates.

Incentives, policies, and campaigns aim to influence behavior.

These approaches operate within existing systems.

They do not address structural misalignment.

The Population Bust is not solely behavioral.

It is systemic.

The Transition to Intervention

The consequences of inversion lead to a central question:

If existing systems no longer support continuity, how can they be redesigned?

This question marks a transition.

From observation to intervention.

The next chapter explores one such intervention.

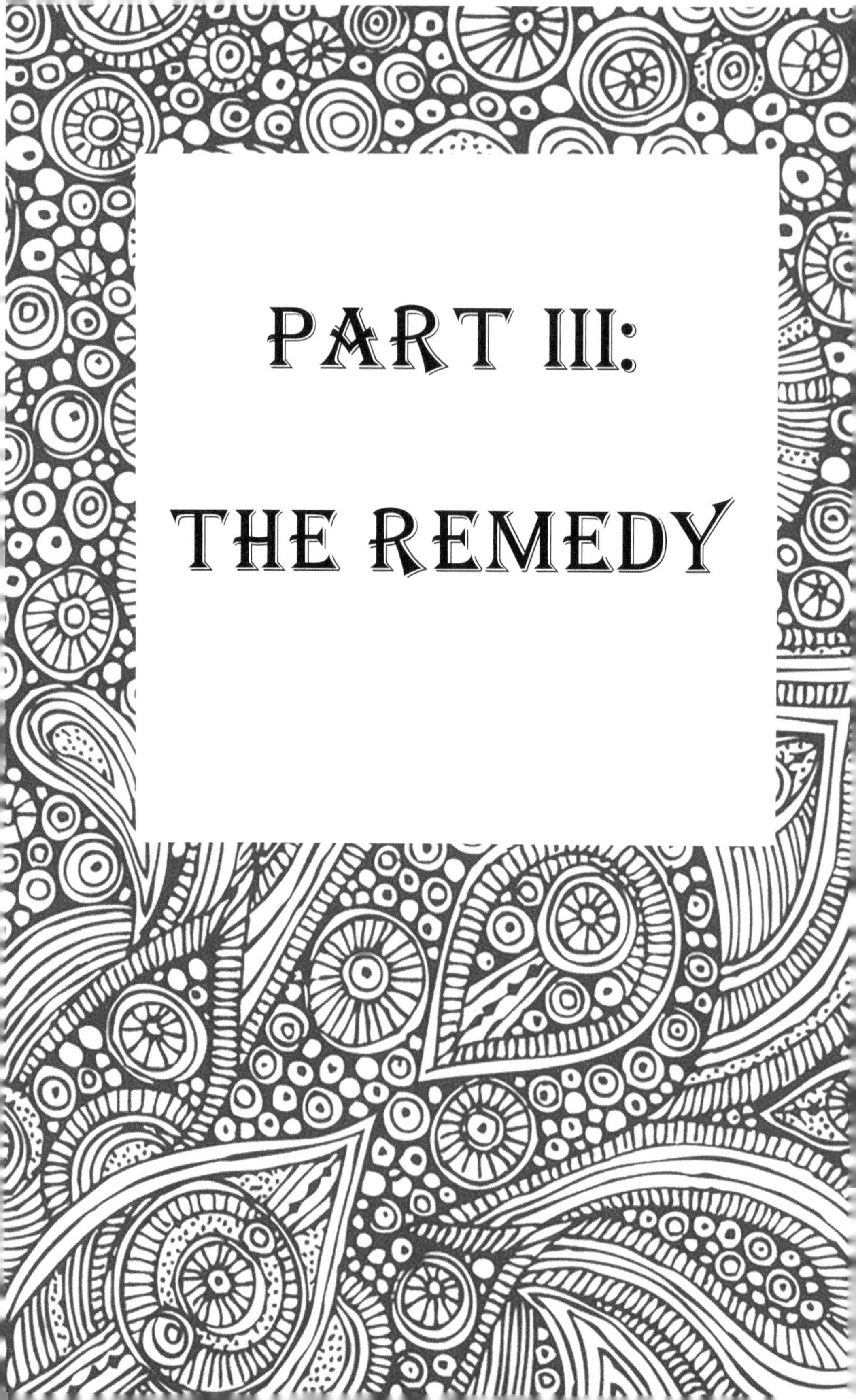

PART III:

THE REMEDY

CHAPTER 5

The Machine Mother – The Tubal Revolution

The preceding chapters have established a structural reality.

The decline in fertility is not incidental. It is not temporary. It is not the result of a single cause that can be corrected through isolated intervention.

It is systemic.

Economic conditions have raised the cost of existence. Social transformations have restructured identity, time, and relationships. Psychological complexity has altered decision-making. Together, these forces have produced a condition in which reproduction is no longer seamlessly integrated into life.

Within this condition, continuity becomes uncertain.

The systems that once ensured generational replacement no longer operate in alignment.

This misalignment raises a fundamental question:

If reproduction can no longer be sustained within its traditional framework, can the framework itself be redefined?

The concept of the Machine Mother emerges from this question.

The Historical Constraint of Biology

For all of human history, reproduction has been bound to biological limitation.

Gestation has required the human body. Pregnancy has imposed physical demands that intersect with time, health, and social participation. These constraints have shaped the structure of life.

They have influenced the timing of relationships, the organization of labor, and the distribution of responsibility.

Even as societies evolved, these constraints remained constant.

Modern systems have changed.

The structure of life has been reorganized around education, career development, and mobility. Time has been extended. Expectations have shifted.

Biology has not adapted to these changes.

This creates tension.

The Machine Mother represents a response to that tension.

Ectogenesis – The Externalization of Gestation

At the center of the Machine Mother lies ectogenesis.

Ectogenesis refers to the development of a fetus outside the human body within a controlled environment.

While complete ectogenesis remains under development, partial forms already exist.

Neonatal technologies have extended the viability of premature infants. Experimental systems have demonstrated the ability to sustain early-stage development outside the womb.

These developments are incremental.

However, their trajectory is clear.

They point toward the possibility of fully externalized gestation.

This possibility represents a transformation.

For the first time, gestation may be separated from the biological body.

The Decoupling of Reproduction from Biology

The most significant implication of ectogenesis is decoupling.

Reproduction becomes independent of biological constraints.

The physical demands of pregnancy are removed.

The limitations of timing are reduced.

Reproduction is no longer bound to a narrow biological window.

It becomes flexible.

This flexibility changes decision-making.

Individuals can align reproduction with economic stability, relational readiness, and long-term planning.

Time becomes adjustable.

Not fixed.

The Reorganization of Time and Life Structure

The decoupling of biology from reproduction alters the structure of life.

The sequence of education, career development, and family formation can be reorganized.

Reproduction no longer competes directly with other life stages.

This reduces conflict.

It allows for integration.

Individuals can pursue stability without sacrificing continuity.

This reorganization addresses one of the central tensions identified in earlier chapters.

The misalignment between modern life and biological reproduction.

The Redistribution of Reproductive Burden

The biological model of reproduction imposes asymmetry.

Pregnancy places physical demands on women.

These demands influence career trajectories, economic participation, and social roles.

Ectogenesis redistributes this burden.

By externalizing gestation, the physical demands are removed.

Both parents participate under similar conditions.

This shift has structural implications.

It alters the framework of parenthood.

Responsibility becomes more balanced.

Decision-making becomes more collaborative.

The Continuum of Reproductive Technology

The Machine Mother is not an isolated innovation.

It is part of a continuum.

Contraception allowed control over conception.

Assisted reproductive technologies enabled conception under specific conditions.

Ectogenesis extends this progression.

It introduces control over gestation.

Each stage expands human influence over reproduction.

The Machine Mother represents the systemic integration of this influence.

Environmental Control and Developmental Precision

Natural gestation occurs within a variable environment.

Nutrition, stress, and external factors influence development.

An artificial gestational system introduces control.

Conditions can be regulated.

Nutrient delivery can be optimized.

Exposure to toxins can be minimized.

Development becomes more predictable.

This introduces the concept of developmental precision.

However, it also raises questions.

The balance between optimization and over-control must be considered.

Ethical Dimensions of Engineered Gestation

The Machine Mother introduces ethical complexity.

Questions emerge regarding control, identity, and access.

To what extent should gestation be engineered?

Who has access to these technologies?

How does externalized gestation affect concepts of parenthood?

These questions are not obstacles.

They are necessary considerations.

Ethical frameworks must evolve alongside technological capability.

Social Integration and Cultural Transition

The adoption of ectogenesis will not occur uniformly.

It will require cultural adaptation.

Concepts of family and reproduction are deeply embedded.

Changes will be gradual.

Acceptance will vary across societies.

Over time, normalization may occur.

As benefits become evident, integration becomes more feasible.

Economic Implications and Accessibility

The economic dimension is significant.

Developing and maintaining artificial gestational systems requires investment.

Cost influences accessibility.

Limited access may create inequality.

Broad access may support stability.

The economic model must align with systemic goals.

Scenario Modeling – Pathways of Adoption

The adoption of the Machine Mother may follow multiple pathways.

Initial use may focus on medical necessity.

Gradual expansion may occur as technology improves.

Normalization may follow.

Alternatively, resistance may slow adoption.

Cultural, ethical, and economic factors will shape these pathways.

Resistance and Psychological Barriers

Technological capability does not guarantee acceptance.

Reproduction is deeply personal.

Changes to its structure may encounter resistance.

Concerns about naturalness, identity, and control may arise.

These concerns must be addressed.

Transparency and understanding are essential.

The Machine Mother as Structural Response

The Machine Mother addresses a structural problem.

It aligns reproduction with modern conditions.

It does not revert to previous systems.

It adapts.

This adaptation expands possibilities.

Integration with Systemic Framework

Ectogenesis alone is insufficient.

It must integrate with economic, social, and environmental systems.

The next chapter explores this integration.

Redefining Continuity

The Machine Mother redefines continuity.

Reproduction becomes intentional.

It becomes part of a designed system.

The Threshold of Transformation

The transition will be gradual.

It will depend on technology, acceptance, and policy.

However, the direction is clear.

Reproduction is entering a new phase.

Toward the Engineered Future

The Machine Mother is not an endpoint.

It is a transition.

It leads to a broader framework.

The Engineered Future.

CHAPTER 6

Sustainability and the Engineered Future

The preceding chapters have led to a singular realization:

Human continuity is no longer a passive outcome of biological and social processes.

It has become an active question.

The systems that once ensured generational replacement—economic structures, social frameworks, and biological rhythms—are no longer aligned. The conditions that supported continuity have evolved, and in doing so, they have disrupted the mechanisms through which continuity was maintained.

The Population Bust is not merely a demographic event.

It is a systemic signal.

It indicates that the structure of human civilization has reached a point at which its foundational assumptions must be reconsidered.

The assumption of automatic continuity can no longer be sustained.

It must be replaced with intentional design.

The Engineered Future represents this transition.

From Evolutionary Momentum to Intentional Architecture

For most of human history, societies evolved in response to external pressures.

Environmental constraints, resource availability, and technological limitations shaped the development of human systems. Adaptation occurred incrementally, often without centralized coordination.

This evolutionary model was sufficient under conditions of continuity.

As long as populations replaced themselves, systems could adjust over time. Growth provided flexibility. Expansion allowed for error.

The Population Bust alters this condition.

When continuity is no longer guaranteed, evolutionary momentum becomes insufficient.

Systems cannot rely on gradual adaptation.

They must be designed.

The shift from evolution to design represents a fundamental change in how civilization operates.

It introduces intention.

It requires coordination.

It demands foresight.

Precision Population – The Principle of Equilibrium

At the center of the Engineered Future lies the concept of Precision Population.

Precision Population is not a fixed number.

It is a principle of alignment.

It recognizes that population size must correspond to the capacity of the systems that support it. Excess creates strain. Deficiency creates instability.

The objective is equilibrium.

This equilibrium is dynamic.

It must adjust to changes in technology, environment, and economic conditions.

Maintaining this balance requires integration across systems.

Reproductive processes must align with economic feasibility. Economic systems must align with infrastructure capacity. Infrastructure must align with environmental sustainability.

Precision Population emerges from this alignment.

It is not imposed.

It is achieved.

Predictive Intelligence and the Management of Complexity

The Engineered Future relies on predictive intelligence.

Modern systems generate vast amounts of data.

Population trends, resource consumption, environmental conditions, and economic activity can be analyzed with increasing precision.

This analysis provides foresight.

Predictive intelligence allows societies to anticipate imbalances before they become critical. It transforms reactive decision-making into proactive planning.

This does not eliminate uncertainty.

However, it reduces unpredictability.

It enables systems to adapt before reaching points of failure.

In the context of population dynamics, predictive intelligence allows for the coordination of reproduction, resource allocation, and infrastructure development.

It provides the informational foundation for design.

Infrastructure as Dynamic System

Infrastructure forms the physical framework of civilization.

It supports transportation, communication, energy distribution, and access to essential resources.

Traditional infrastructure is designed for growth.

Expansion is assumed.

The Engineered Future requires a different model.

Infrastructure must be dynamic.

It must adapt to changing conditions.

It must function under both growth and decline.

This requires flexibility.

Systems must be scalable, modular, and resilient.

They must adjust to population levels rather than assume expansion.

The Subterranean Arterial Network (SAN) – Structural Efficiency

The Subterranean Arterial Network represents a model of integrated infrastructure.

By relocating key systems underground, SAN reduces environmental exposure and optimizes resource distribution.

Temperature stability reduces energy demand.

Protection from external disruption increases reliability.

Surface space can be preserved for ecological and social use.

This model reflects a broader principle:

Infrastructure can be designed to enhance sustainability rather than merely support growth.

The SHIELD Framework – Environmental Stabilization

Complementing SAN is the SHIELD system.

This framework focuses on environmental regulation.

Through continuous monitoring and adaptive response mechanisms, SHIELD mitigates environmental variability.

Air quality, temperature, and exposure to pollutants can be managed.

The goal is not to eliminate variability.

It is to reduce its impact.

This creates a stable environment for human systems.

Closed-Loop Civilization – Beyond Linear Consumption

Traditional systems operate on linear models.

Resources are extracted, used, and discarded.

This model is efficient in the short term but unsustainable over time.

The Engineered Future adopts a closed-loop model.

Resources are reused.

Waste is minimized.

Systems maintain equilibrium.

This principle extends beyond material resources.

It includes population dynamics.

Population must align with resource availability.

This alignment ensures long-term sustainability.

Integration of Reproductive Systems

Reproductive systems must be integrated into this framework.

Technologies such as ectogenesis provide tools for managing population.

However, they must align with other systems.

Population size must correspond to infrastructure capacity, economic stability, and environmental limits.

This integration transforms reproduction into a coordinated system.

Continuity becomes intentional.

Economic Alignment and Stability

Economic systems must support sustainability.

This requires stability.

Volatility must be reduced.

Long-term planning must be prioritized.

Population stability contributes to economic predictability.

Predictable demand supports efficient production.

Balanced labor markets support productivity.

Economic systems reinforce continuity.

Social Systems and Human Experience

Sustainability is not solely technical.

It is social.

Human experience must be considered.

Systems must support identity, community, and purpose.

The Engineered Future must align individual fulfillment with collective sustainability.

This requires rethinking social structures.

Communities must provide support.

Responsibilities must be distributed.

Continuity must coexist with autonomy.

Governance and Ethical Design

Intentional design introduces governance.

Decisions about population, infrastructure, and resources must be guided by principles.

Equity ensures fairness.

Sustainability ensures longevity.

Transparency ensures trust.

Governance structures must balance control and freedom.

They must enable coordination without imposing rigidity.

Scenario Modeling – Possible Futures

The Engineered Future is not singular.

Multiple pathways exist.

One scenario involves gradual integration.

Technologies are adopted incrementally.

Systems adjust over time.

Another scenario involves rapid transition.

Structural pressures accelerate change.

Systems are redesigned quickly.

A third scenario involves resistance.

Adoption is delayed.

Systems degrade before adaptation occurs.

Each scenario reflects different responses.

The outcome depends on decision-making.

The Psychological Dimension of Design

Design is not purely technical.

It is psychological.

Individuals must accept new frameworks.

Trust must be established.

Perceptions must shift.

The success of the Engineered Future depends on alignment between systems and human perception.

Legacy and Intergenerational Continuity

At its core, sustainability is about continuity.

Human civilization exists across generations.

Each generation inherits systems.

Each generation modifies them.

Legacy is the transmission of these systems.

The Population Bust threatens this transmission.

The Engineered Future restores it.

It ensures that continuity is supported.

The Ethical Boundary of Control

Design introduces control.

Control must be balanced.

Excessive control risks rigidity.

Insufficient control risks instability.

The ethical boundary lies in alignment.

Systems must support without constraining.

The Emergence of a New Paradigm

The Engineered Future represents a paradigm shift.

From growth to balance.

From assumption to intention.

From evolution to design.

This shift redefines civilization.

Continuity as Constructed Outcome

Continuity is no longer automatic.

It is constructed.

Population, infrastructure, economy, and environment must align.

This alignment produces stability.

Stability sustains civilization.

The Open Horizon

The future is not predetermined.

It is shaped by decisions.

The Engineered Future provides a framework.

It offers a path.

The direction depends on choice.

Final Synthesis

The Population Bust marks a transition.

The response to this transition defines the future.

The Engineered Future offers a means of alignment.

It transforms continuity from assumption into design.

CHAPTER 7

The Choice – The Point of Divergence

All systems, when pushed beyond alignment, reach a point at which continuation is no longer automatic.

They reach a threshold.

At that threshold, the trajectory of the future is no longer determined by momentum alone. It becomes dependent on response.

Human civilization now stands at such a threshold.

The Population Bust is not an isolated phenomenon. It is the cumulative result of economic, social, biological, and psychological transformations that have unfolded over time. These transformations have altered the conditions under which continuity is maintained.

The systems that once supported generational replacement no longer operate in alignment.

The result is not immediate collapse.

It is gradual divergence.

This divergence creates a point of decision.

This is the Choice.

The Nature of Civilizational Choice

The concept of choice, when applied to individuals, implies intention and agency.

When applied to civilizations, it becomes more complex.

Civilizations do not decide in a single moment.

They move through patterns of response.

Policies are implemented. Systems are adjusted. Behaviors evolve. Technologies are adopted or resisted.

These responses, taken together, form a trajectory.

The Choice is not a singular decision.

It is a directional outcome.

It emerges from the accumulation of responses across systems.

At the present moment, multiple trajectories are possible.

Each leads to a different future.

The Illusion of Non-Choice

One of the defining characteristics of systemic transitions is the illusion that no decision is being made.

When change occurs gradually, it can be mistaken for continuity.

Systems continue to function.

Institutions remain in place.

Daily life proceeds without interruption.

This creates the perception that no action is required.

However, inaction is not neutral.

It is a form of choice.

When systems are misaligned, maintaining the status quo allows the misalignment to persist.

Over time, this persistence produces outcome.

The absence of intervention becomes a trajectory.

Pathway One – Passive Continuation

The first trajectory may be described as passive continuation.

In this pathway, existing systems remain largely unchanged.

Economic structures continue to operate as they are. Social frameworks evolve incrementally. Technological developments proceed without coordinated integration.

Reproduction remains embedded within current conditions.

No systemic realignment occurs.

At first, the effects are subtle.

Population decline continues gradually.

Aging populations increase.

Economic systems adjust to accommodate demographic shifts.

Infrastructure operates with increasing inefficiency.

Over time, these adjustments accumulate.

The dependency ratio worsens.

Workforce contraction reduces economic output.

Public systems experience sustained pressure.

This pathway does not produce immediate collapse.

It produces gradual contraction.

Societies become smaller, older, and less dynamic.

Innovation slows.

Risk tolerance decreases.

Cultural orientation shifts toward preservation rather than expansion.

This trajectory may stabilize at lower population levels.

However, the process of reaching that point is marked by prolonged adjustment.

The cost is distributed over time.

It is absorbed gradually.

The Psychology of Passive Decline

Passive continuation is not driven by explicit intent.

It emerges from psychological patterns.

Individuals adapt to conditions as they are.

Institutions respond to immediate pressures rather than long-term trends.

Decision-making prioritizes short-term stability.

The absence of urgency reinforces inaction.

The gradual nature of demographic change contributes to this pattern.

Without visible disruption, the need for intervention is not widely perceived.

The system continues.

The trajectory remains unchanged.

Pathway Two – Policy-Based Correction

The second trajectory involves targeted intervention.

In this pathway, governments and institutions recognize the decline in fertility and attempt to address it through policy.

Financial incentives are introduced.

Childcare support is expanded.

Parental leave policies are adjusted.

Housing initiatives are implemented.

These measures aim to reduce the cost of parenthood and encourage higher birth rates.

In some cases, they produce measurable effects.

Fertility rates may increase temporarily.

The pace of decline may slow.

However, these effects are often limited.

They address surface-level conditions rather than underlying structures.

The Structural Impasse remains.

Economic pressures persist.

Social timelines remain extended.

Psychological complexity continues to influence decision-making.

As a result, policy-based correction produces partial stabilization.

It mitigates decline.

It does not reverse it.

The Limits of Incentive-Based Systems

Incentives operate within existing frameworks.

They influence behavior by adjusting cost and benefit.

However, when the underlying system is misaligned, incentives have limited impact.

Reproduction is not solely a financial decision.

It is shaped by time, identity, and structural conditions.

Policies that address cost without addressing alignment produce constrained outcomes.

The system remains.

The trajectory shifts slightly.

The underlying dynamic persists.

Pathway Three – Technological Integration

The third trajectory introduces technological adaptation.

In this pathway, advancements in reproductive technology are integrated into the system.

Ectogenesis becomes viable.

The Machine Mother transitions from concept to implementation.

Reproduction is no longer constrained by biological timelines.

It becomes flexible.

This flexibility alters decision-making.

Individuals can align reproduction with modern life structures.

The conflict between career and parenthood is reduced.

Timing becomes adjustable.

This pathway addresses one dimension of the Structural Impasse.

It resolves the biological constraint.

However, it does not automatically resolve economic or social conditions.

Integration is required.

Technological capability must align with access, policy, and cultural acceptance.

When integration occurs, this pathway can stabilize population dynamics.

Continuity becomes more feasible.

The Adoption Curve and Social Acceptance

Technological integration does not occur uniformly.

Adoption follows patterns.

Early adoption is limited.

Acceptance grows gradually.

Resistance emerges alongside acceptance.

Cultural, ethical, and psychological factors influence the rate of integration.

The trajectory depends on these factors.

In some contexts, adoption may accelerate.

In others, it may remain constrained.

The outcome is variable.

Pathway Four – The Engineered Future

The fourth trajectory represents systemic realignment.

In this pathway, population, infrastructure, economy, and environment are intentionally aligned.

Reproduction is integrated into a broader framework.

Technological systems support biological processes.

Economic systems support family formation.

Infrastructure adapts to population dynamics.

Environmental systems maintain sustainability.

This pathway reflects design.

Continuity is not left to chance.

It is constructed.

Precision Population becomes achievable.

Systems operate in coordination.

Stability is maintained.

This trajectory represents the most comprehensive response.

It requires the highest level of coordination.

It introduces complexity.

It also offers the greatest potential for long-term sustainability.

The Conditions for Transition

The transition between pathways is not immediate.

It requires recognition.

It requires alignment of systems.

It requires coordination across domains.

The feasibility of each pathway depends on multiple factors:

Technological development

Economic capacity

Cultural acceptance

Political will

These factors interact.

They shape the trajectory.

The Role of Time in Decision

Time influences the Choice.

The longer misalignment persists, the greater the structural pressure.

Delayed response reduces available options.

Early recognition expands them.

This dynamic creates urgency.

Not immediate urgency.

But structural urgency.

The Ethical Dimension of Choice

Each pathway carries ethical implications.

Passive continuation prioritizes autonomy but risks long-term instability.

Policy-based correction attempts to balance support and freedom.

Technological integration introduces questions of access and identity.

The Engineered Future requires governance and coordination.

Ethical frameworks must guide these decisions.

Equity, sustainability, and autonomy must be considered.

The Cost of Inaction

Inaction is not neutral.

It allows existing trajectories to continue.

The cost of inaction is cumulative.

It manifests in economic strain, infrastructure inefficiency, and demographic imbalance.

It is not immediate.

It is gradual.

Yet it is real.

The Opportunity of Alignment

The Choice is not solely about avoiding decline.

It is about creating alignment.

Alignment between systems.

Alignment between conditions and outcomes.

Alignment between individual decisions and collective sustainability.

This alignment represents opportunity.

The Human Dimension

At its core, the Choice is human.

It affects how individuals live, decide, and plan for the future.

It shapes the conditions under which families form.

It influences the continuity of generations.

This dimension must remain central.

Systems exist to support human life.

Not replace it.

The Open Decision

The Choice remains open.

No single trajectory is predetermined.

Multiple pathways exist.

Each reflects different responses.

The outcome depends on collective action.

The Point of Divergence

The present moment represents a point of divergence.

The direction taken from this point will define the structure of the future.

The Population Bust is not the end of continuity.

It is the moment at which continuity must be reconsidered.

Toward the Final Reflection

The Choice leads to reflection.

The implications extend beyond systems.

They touch on meaning, purpose, and legacy.

The final chapter addresses these dimensions.

CONCLUSION

Zero Hour – The Moment That Defines Continuity

There are moments in the trajectory of civilization that do not announce themselves with noise.

They do not arrive with immediate disruption or visible collapse.

They emerge quietly.

They take shape over time.

They are revealed not through what is present, but through what is absent.

The Population Bust is such a moment.

It does not demand attention in the way that crises typically do. It does not disrupt daily life with sudden force. It does not produce immediate alarm.

Instead, it unfolds gradually.

It manifests in empty classrooms, in aging communities, in the subtle absence of new generations.

It is not a collapse.

It is a quiet divergence.

And yet, its significance is profound.

Because what is at stake is not a system.

It is continuity itself.

The End of Assumption

For most of human history, continuity required no justification.

It was embedded within the structure of life.

Generations followed one another without intervention.

Reproduction was not a question.

It was a consequence.

This condition shaped how societies understood themselves.

Economic systems assumed expanding populations.

Infrastructure was designed for growth.

Cultural narratives emphasized lineage and inheritance.

Continuity was not managed.

It was assumed.

That assumption no longer holds.

The systems that once supported automatic replacement have evolved.

Economic pressures have increased the cost of existence.

Social transformations have restructured identity and time.

Psychological complexity has altered decision-making.

Biological constraints remain unchanged.

Together, these forces have produced a condition in which continuity is no longer guaranteed.

It has become conditional.

The Silent Accumulation

The defining characteristic of this transition is its silence.

Unlike visible crises, which demand immediate response, the Population Bust operates through accumulation.

Its effects build gradually.

They are distributed across time.

They do not trigger urgency.

This absence of urgency creates risk.

When change is not perceived as immediate, response is delayed.

When response is delayed, structural conditions persist.

The trajectory continues.

The cost accumulates.

This is the nature of Zero Hour.

It is not the moment of collapse.

It is the moment before it becomes irreversible.

The System Revealed

The analysis presented in this work has revealed a system.

Economic Sterilization constrains the feasibility of reproduction.

The Structural Impasse disrupts its integration into life.

Demographic inversion reshapes the architecture of society.

Technological adaptation introduces new possibilities.

The Engineered Future offers a framework for alignment.

These are not isolated observations.

They are components of a single system.

The decline in fertility is not an anomaly.

It is the outcome of that system.

Understanding this is essential.

Because systems cannot be addressed through isolated measures.

They require alignment.

The Cost of Inaction

Inaction is often misunderstood as neutrality.

It is not.

Inaction allows existing trajectories to continue.

It reinforces current conditions.

It sustains misalignment.

The cost of inaction is cumulative.

It manifests in demographic imbalance, economic strain, and structural inefficiency.

It unfolds gradually.

It is absorbed across generations.

It is not immediately visible.

Yet it is real.

The absence of immediate crisis does not imply the absence of consequence.

It delays recognition.

The Nature of Response

The response to the Population Bust is not singular.

It exists across a spectrum.

Passive continuation allows decline to proceed.

Policy-based correction attempts to mitigate its effects.

Technological integration introduces adaptation.

Systemic alignment redefines the framework.

Each response reflects a different level of engagement.

Each produces a different outcome.

The future is shaped by these responses.

The Responsibility of Design

If continuity is no longer automatic, it must be supported.

This introduces responsibility.

The responsibility to understand systems.

The responsibility to align them.

The responsibility to act.

Design becomes central.

Not as control.

But as coordination.

The Engineered Future represents this approach.

It aligns population, infrastructure, economy, and environment.

It transforms continuity from assumption into outcome.

The Human Question

Amid systems and structures, a fundamental question remains:
What does continuity mean?
It is not merely the persistence of numbers.
It is the continuation of experience.
It is the transmission of knowledge, identity, and possibility.
It is the presence of future generations.

The Population Bust challenges this presence.

It introduces the possibility of absence.

This possibility is not abstract.

It is measurable.

It is observable.

It is unfolding.

The Threshold of Choice

Zero Hour represents a threshold.
A point at which continuation is no longer determined by momentum.
It is determined by choice.
Not a singular decision.
But a collective trajectory.
The pathways are clear.
Passive decline.
Partial adaptation.
Technological integration.
Systemic alignment.
Each leads to a different future.

The Future as Constructed Reality

The future is no longer inherited.

It is constructed.

The conditions that shape it are influenced by decisions made in the present.

These decisions affect population dynamics, economic systems, and environmental sustainability.

They define the structure of continuity.

The shift from inheritance to construction is fundamental.

It changes how humanity relates to time.

The Alignment Imperative

The central requirement of the future is alignment.

Alignment between systems.

Alignment between conditions and outcomes.

Alignment between individual decisions and collective sustainability.

Without alignment, systems diverge.

With alignment, they stabilize.

This is the imperative of the Engineered Future.

The Final Recognition

The Empty Cradle is not merely a symbol.

It is a signal.

A signal that the conditions of continuity have changed.

A signal that adaptation is required.

A signal that the future is no longer automatic.

This recognition defines Zero Hour.

The Last Question

At this threshold, a final question emerges:

Will continuity remain an assumption?

Or will it become a design?

The answer to this question will define the trajectory of humanity.

Closing Reflection

The Population Bust is not the end of civilization.

It is a transition.

A moment at which long-standing assumptions must be reconsidered.

A moment at which systems must be realigned.

A moment at which the future must be understood differently.

Zero Hour is not defined by what has already occurred.

It is defined by what follows.

The direction is not fixed.

It remains open.

And in that openness lies both risk and possibility.

The future will not emerge on its own.

It will be shaped.

And in that shaping, humanity defines its continuation.

AFTERWORD

The Child Not Yet Born

Every civilization is built not only for those who live within it, but for those who are expected to come after.

This expectation is so deeply embedded in human life that it is rarely examined. Roads are built with future use in mind. Schools are established for future students. Laws, institutions, and traditions are preserved because they are assumed to have inheritors. Even the language of planning—legacy, continuity, development, investment—carries within it an unspoken assumption that there will be a next generation to receive what has been built.

The child not yet born is present in nearly every serious human undertaking.

And yet, in the modern world, that child has become increasingly abstract.

Not because human beings no longer care about the future, but because the relationship between the present and the future has become more uncertain. The conditions under which new life once emerged as a natural continuation of life itself have changed. The path from adulthood to parenthood is no longer structurally embedded. It is negotiated, delayed, or set aside within systems that increasingly reward flexibility, specialization, and self-preservation.

The child not yet born does not disappear through rejection alone.

That child disappears through postponement, through structural incompatibility, through the quiet accumulation of pressures that make continuity more difficult to choose.

This is why the issue examined in this book is not merely demographic.

It is moral in the broadest and most civilizational sense.

Not moral in the language of blame, but moral in the language of obligation. What, if anything, do the living owe to those who do not yet exist? What responsibility does a civilization bear toward its own continuation? And what does it mean when the systems of that civilization no longer support the emergence of those who are meant to inherit it?

These questions do not have simple answers.

But they cannot be ignored.

The child not yet born represents possibility. Not just biological possibility, but cultural, intellectual, and civilizational possibility. Every generation carries within it people who will think thoughts not yet thought, build structures not yet built, solve problems not yet understood, and reframe the world in ways the present cannot foresee. The future is not merely a continuation of current systems. It is also their correction, expansion, and renewal.

A society that loses confidence in the emergence of future generations loses something deeper than its fertility rate.

It loses part of its orientation toward time.

It begins to think in terms of maintenance rather than inheritance, management rather than renewal, endurance rather than aspiration. This shift is subtle, but it affects everything. It affects how institutions behave, how communities plan, and how individuals imagine their place in history.

To care about the child not yet born is to care about more than population.

It is to care about whether the future remains open.

That openness is not guaranteed. It depends on conditions— economic conditions, social conditions, environmental conditions, and psychological conditions. It depends on whether systems remain capable of supporting continuity. It depends on whether the burdens of existence have grown so heavy that the act of bringing new life into the world becomes structurally discouraged.

This is why continuity can no longer be treated as automatic.

If the child not yet born is to remain more than an abstraction, then the systems surrounding human life must be made capable of receiving that child.

This requires more than sentiment.

It requires alignment.

It requires societies to ask whether the structures they have built are capable of sustaining not only the present, but the future. It requires a reconsideration of what progress means if it expands autonomy while weakening continuity. It requires the courage to confront the possibility that modern civilization may be materially advanced while demographically fragile.

And yet, this recognition should not produce despair.

It should produce seriousness.

The Population Bust is not meaningful because it signifies an end. It is meaningful because it reveals a condition that can still be addressed. The child not yet born remains possible. The future remains open. The systems of the present are not fixed beyond revision. They can be examined, challenged, and redesigned.

That is the deeper argument of this book.

Not merely that humanity faces a demographic transition, but that it stands at a moment when continuity must become intentional. If previous generations could assume inheritance, the present generation must build for it consciously.

The child not yet born is not simply waiting to arrive.

That child depends on the choices of the living.

On the structures they preserve.

On the systems they reform.

On the degree to which they are willing to recognize that continuity is not a background condition of civilization, but one of its highest responsibilities.

In the end, every serious society must answer a question that is both simple and profound:

What kind of future is it preparing to hand forward?

The answer is not found in rhetoric, nor in statistics alone.

It is found in whether that future can still receive new life.

Whether it can still make room for inheritance.

Whether it can still sustain the child not yet born.

AUTHOR'S NOTE

This book was written in response to a quiet but consequential reality: humanity is entering a period in which continuity can no longer be assumed.

For generations, the dominant population narrative centered on expansion. The question was how to manage growth, how to distribute resources, and how to sustain the environmental and economic systems required by an increasing number of people. Yet beneath that widely accepted narrative, a different pattern began to emerge. Birth rates declined. Populations aged. The structure of society began to shift, not through sudden disruption, but through gradual change.

What makes this transformation significant is not only its scale, but its silence.

The Population Bust does not announce itself dramatically. It appears in declining enrollment, delayed family formation, rising dependency ratios, and the growing distance between the biological framework of reproduction and the social systems of modern life. It is a structural issue disguised as a private one. It is experienced through individual decisions, but produced by systemic conditions.

This manuscript was written to examine that condition in full.

It is not intended as an argument against progress. On the contrary, many of the forces contributing to declining fertility—expanded education, economic mobility, technological advancement, and individual autonomy—represent genuine achievements of modern civilization. But every achievement carries structural consequences. The question is whether the systems surrounding those achievements have evolved enough to preserve continuity.

The answer, increasingly, appears to be no.

This book approaches the issue from multiple angles because no single angle is sufficient. Fertility decline is not only economic. It is not only social. It is not only biological. It is not only psychological. It is the product of interaction among these forces. To treat it as a single-variable problem is to misunderstand its nature.

The purpose of this work, therefore, is not merely to describe decline, but to reframe the conversation around it.

The central premise of this book is that continuity has moved from assumption to design. Human societies can no longer rely on inherited demographic momentum to sustain themselves indefinitely. The systems that once supported generational replacement must now be examined, challenged, and, where necessary, redesigned.

That redesign may involve policy. It may involve infrastructure. It may involve technological adaptation. It may require cultural and ethical reconsideration. Whatever form it takes, the essential point remains the same: if continuity is to be maintained, it must become intentional.

This manuscript is both analytical and speculative. It is analytical in its diagnosis of the pressures shaping the Population Bust. It is speculative in its exploration of how humanity may respond. The discussion of ectogenesis, precision population, and engineered systems is not offered as fantasy, nor as inevitability, but as part of a broader effort to think beyond inherited assumptions.

Civilizations change when the conditions that sustain them change.

We are living through such a change now.

If this book succeeds in any respect, it will not be because it offers a single definitive answer. It will be because it broadens the field of inquiry. It will be because it invites readers to see the Population Bust not as an isolated demographic curiosity, but as a civilizational threshold.

The future will not be shaped by awareness alone.

But awareness is where it begins.

APPENDICES

APPENDIX A

Key Definitions and Core Concepts

This appendix provides clear definitions of the primary concepts introduced in this work to ensure consistency of interpretation.

Population Bust

A sustained decline in fertility rates below replacement level, resulting in long-term population contraction.

Replacement Rate

The fertility level required to maintain a stable population across generations, typically estimated at 2.1 children per woman.

Apex

The transition phase at which population growth shifts from expansion to decline.

Structural Impasse

A systemic misalignment between modern social structures and the biological realities of reproduction.

Economic Sterilization

A condition in which economic factors function as constraints on reproduction despite biological capacity.

The Great Quiet

The gradual and largely unnoticed decline in birth rates across developed and developing societies.

APPENDIX B

The Precision Population Model

The Precision Population model represents a framework for maintaining equilibrium between population size and system capacity.

Core Principles:

1. **Equilibrium Over Expansion**

   Population should align with available resources and infrastructure capacity.

2. **Dynamic Adjustment**

   Population levels should be responsive to changing environmental, economic, and technological conditions.

3. **System Integration**

   Reproductive systems, economic structures, and infrastructure must operate cohesively.

Model Implications:

- Prevents overpopulation strain

- Avoids demographic collapse

- Supports long-term sustainability

APPENDIX C

The Subterranean Arterial Network (SAN) – Concept Overview

The Subterranean Arterial Network (SAN) is a conceptual infrastructure model designed to optimize efficiency and sustainability.

Key Features:

- Underground transportation and logistics systems
- Climate-stable environments
- Reduced surface congestion
- Energy-efficient operations

Functional Benefits:

- Increased infrastructure resilience
- Reduced environmental exposure
- Improved resource distribution

APPENDIX D

The SHIELD Environmental Framework

The SHIELD system represents an integrated environmental management framework.

Objectives:

- Stabilize environmental conditions
- Reduce exposure to pollutants
- Maintain sustainable living environments

Components:

- Air quality monitoring systems
- Climate regulation technologies
- Environmental feedback loops

Outcome:

A controlled yet adaptive environment supporting both human and ecological systems.

APPENDIX E

Ectogenesis and the Machine Mother – Technical Overview

This appendix outlines the foundational concepts behind ectogenesis.

Definition:

Ectogenesis is the development of a fetus outside the human body within an artificial gestational system.

Current State:

- Partial ectogenesis (neonatal support systems)
- Experimental artificial womb technologies

Future Potential:

- Full external gestation
- Reduced biological constraints
- Alignment with modern life structures

Considerations:

- Ethical frameworks
- Accessibility
- Regulatory systems

APPENDIX F

System Interaction Map

The Population Bust is not caused by a single factor, but by the interaction of multiple systems.

Core Systems:
- Economic System
- Social System
- Biological System
- Technological System
- Environmental System

Interaction Framework:

Economic Pressure → Delayed Stability → Delayed Reproduction

Social Shift → Identity Expansion → Reduced Default Parenthood

Biological Limits → Fixed Fertility Window → Timing Conflict

Technological Intervention → Reproductive Adaptation

Environmental Constraints → Sustainability Pressure

Outcome:

A systemic condition requiring integrated solutions.

APPENDIX G

Scenario Pathways for the Future

The future of population dynamics may follow multiple pathways:

1. Passive Decline

- Continued fertility reduction
- Aging populations
- Economic contraction

2. Policy-Based Stabilization

- Incentives and reforms
- Limited effectiveness

3. Technological Integration

- Adoption of ectogenesis
- Alignment with modern systems

4. Engineered Future

- Full system integration
- Precision Population achieved

APPENDIX H

Ethical Framework for the Engineered Future

Any transition toward an engineered system must be guided by ethical principles:

Equity

Access to systems must be fair and inclusive.

Sustainability

Systems must support long-term viability.

Autonomy

Individual choice must be preserved.

Transparency

Decision-making processes must be clear and accountable.